U0392556

人类能更人性化吗?

Can Humans Become More Humane?

Gunter Pauli

[比] 冈特·鲍利 著

[哥伦] 凯瑟琳娜·巴赫 绘

高青 译

上海远东出版社

丛书编委会

主　任：田成川

副主任：闫世东　林　玉

委　员：李原原　祝真旭　曾红鹰　靳增江　史国鹏
　　　　梁雅丽　孟小红　郑循如　陈　卫　任泽林
　　　　薛　梅　朱智翔　柳志清　冯　缨　齐晓江
　　　　朱习文　毕春萍　彭　勇

特别感谢以下热心人士对童书工作的支持：

匡志强　宋小华　解　东　厉　云　李　婧　庞英元
李　阳　梁婧婧　刘　丹　冯家宝　熊彩虹　罗淑怡
旷　婉　王靖雯　廖清州　王怡然　王　征　邵　杰
陈强林　陈　果　罗　佳　闫　艳　谢　露　张修博
陈梦竹　刘　灿　李　丹　郭　雯　戴　虹

目录

Contents

黑暗中，一只蟑螂在屋子里爬行，他在寻找藏身之地。在冰箱后面、他的窝附近，一个既靠近食物又安全温暖的地方，他遇到了一只老鼠。

"你好！"蟑螂说。"你介意我在这儿停一会儿吗？"

A cockroach runs through a house in the dark, looking for a place to hide. He meets a rat, near his home behind the refrigerator, a safe and warm spot that is close to food.
"Hello!" says the cockroach. "Do you mind me parking off here for a while?"

一只蟑螂在屋子里爬行······

A cockroach runs through a house ...

......你自己收拾干净就行。

... clean up after yourself.

"没问题。"老鼠回答道，"只要你自己收拾干净就行。"

　　"自己收拾干净？你什么意思？我想我会是你见过的最干净的生物之一！我会清理干净并吃掉人们留下的所有废弃物。"

"No problem at all," responds the rat, "provided that you clean up after yourself."

"Clean up after myself? What do you mean? I am one of the cleanest beings you'll ever come across! I clean up after others – eating all the waste that people leave behind."

"可人们还是一看到你就要杀你，对吧？我也是同样境况。"

"杀我？哦，他们的确想这么做。为了毒害我，他们把那些我们蟑螂吃了会死的药球到处撒放。"

"欢迎来到'友善的'人类主宰的世界。"

"And yet they kill you when they see you, right? The same applies to me."

"Kill me? Oh, they do try. To poison me – putting those pellets down everywhere so we roaches will die after eating it."

"Welcome to the world dominated by **Mankind**."

为了毒害我……

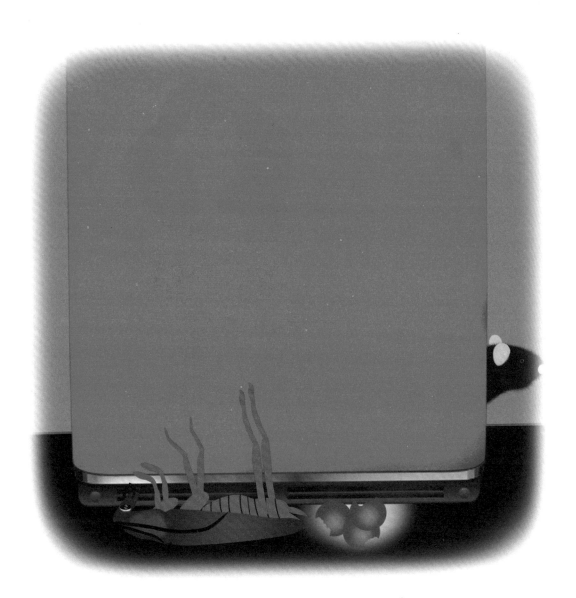

To poison me ...

谁说他们是友善的?

Who said they were kind?

"友善的人类！谁说他们是友善的？"

"嗯，以前他们不是这样的。"老鼠叹了口气说。"记得我奶奶告诉过我，她从老祖那里听到的人类刚开始由四肢行走变成两条腿走路的故事。"

"Mankind! Who said they were **kind**?"

"Well, it never used to be like this." The rat says with a sigh. "I remember my grandma told me stories she heard from her ancestors. About the way these human beings started behaving once they walked on two legs instead of four."

"我很清楚的是，人类根本不讲人道。只要是不喜欢或不理解的东西，他们就杀死！"蟑螂说。

"我相信很久以前他们要友善得多。他们出去觅食和狩猎，回家后和家族的其他人分享这一切。由此获得族人的爱戴和尊重，并以这种方式建立社区。"

"此后发生了什么事？"

"What is very clear to me, is that humans are everything but humane. Whatever they do not like or understand, they kill!" Cockroach says.

"I believe they were much kinder a long time ago. When they went out to forage and hunt, they would get home and share it all with the others of their clan. Gaining love and respect in return, and in this way building communities."

"And then what happened?"

......很久以前他们要友善得多。

... they were much kinder a long time ago.

他们的大脑停止进化了。

Their brain stopped growing.

"他们的大脑停止进化了。"

"什么！他们的大脑停止生长？"

"的确是的，他们并没有好好享受生活，享受他们繁荣的小社区，反而由于别人拥有自己没有的东西而开始互相嫉妒。"

"Their brain stopped growing."

"What! Their brain stopped growing?"

"Indeed, instead of enjoying life, and their thriving small communities, they started to envy each other for what the other one had."

"甚至在最初没人有很多东西之前？"

"唉，一旦脑子里有了喜欢什么或讨厌什么的倾向，人类就产生了许多消极的想法——他们开始担心。"

"担心？每个人当然都有权利担心！我就担心我的生存。"

"Even when before that no one had anything much to start with?"

"Well, as soon as the mind has the option to be in favour or against something, those humans formed a lot of negative opinions – and they started worrying."

"Worrying? Surely everyone has the right to do that! I am worried about my survival."

我就担心我的生存。

I am worried about my survival.

……他们开始对一切进行分析。

... they started analysing everything.

"你看，他们生存了下来并确保每个人都有食物，但不知怎的，他们开始对一切进行分析。"

"是的，我认为如果你过度分析数据，你就不会再有梦想，也会错过许多机遇。"

"没错。如果你看不到可以寻找到水、食物和庇护所的机会，那么你就会担心，可能永远都不会快乐。"

"You see, they survived and secured food that they could all share, but then somehow they started analysing everything."

"Yes, I believe if you go overboard – analysing data too much – you no longer have dreams, and many opportunities are lost."

"True. If you do not see opportunities to find water, food and shelter, then you worry and can never be happy."

"更糟糕的是，如果你只关注这些问题，你就没有机会去畅想等待着我们的美好未来。真希望人类很快就能再次找到梦想。"蟑螂说。

"想象一下，在一个人类真正变得人道的世界里，我们将拥有多么美好的未来！"

……这仅仅是开始！……

"What makes it worse, is if you only focus on the problems, you stand no chance of imagining the grand future that awaits us all. Let's hope humans soon can dream again," Cockroach says.

"Just imagine what a great future we will have in a world where humans are truly humane!"

... AND IT HAS ONLY JUST BEGUN! ...

……这仅仅是开始！……

.. AND IT HAS ONLY JUST BEGUN! ..

Did You Know?

你知道吗?

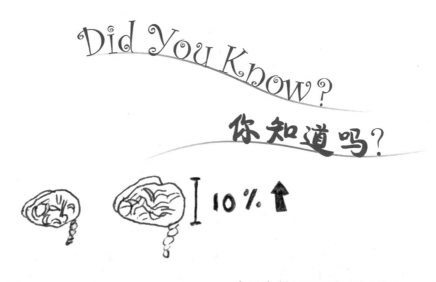

Over millions of years the human brain expanded from 500 cm³ to 1,500 cm³. The largest scull with the biggest brain belongs to the Neanderthal, which is at least 10% bigger than the average size brain at present.

经过数百万年的进化，人脑的大小从 500 立方厘米扩大到了 1500 立方厘米。尼安德特人的脑容量最大，至少比目前平均人脑大 10%。

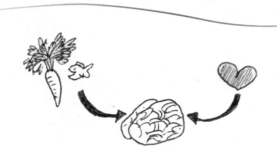

That tremendous brain growth stage is the subject of a great deal of scientific debate. It is clear that a change of food, including seafood, had a major effect. Other researchers defend the hypothesis that it was the development of emotions like loving and caring that caused the brain to expand.

如此巨大的大脑发育进程是大量科学辩论的主题。目前已知包括海鲜在内的食物摄入变化对此有很大影响。但有研究者认为，是爱和关怀等情感的发展导致了大脑扩展。

The greatest invention of all time could well have been the bag, which permitted the hunter-gatherer to collect and bring home 15 times more food than the individual could eat. Returning home with the bounty was therefore met with lots of joy.

有史以来最伟大的发明可能是包，它让狩猎者能把一个人食量15倍多的食物带回家。带着丰厚的礼物回家会收获很多快乐。

In life there are always many problems to resolve, but in the end it is the state of mind of a person that will determine the ability to transform a hard reality (especially one that merits critique), into a vision to improve the situation, and do better.

人生总有许多问题需要解决，但说到底是一个人的心态决定了他能否改善自己面临的困境（尤其是受到指责时），转危为机，让其变得更好。

Cockroaches are omnivorous scavengers and consume any organic matter. They prefer to eat sweet things, meat and starches, but would also eat hair, paper and wood.

蟑螂是杂食性食腐动物，吞食任何有机物质。它们喜欢吃甜的东西、肉类和淀粉，但也会吃头发、纸张和木材。

Rats in Norway prefer meat scraps. Roof rats in the tropics prefer fruit. In urban areas they will eat pet food and raid garbage cans, and there are rats that will even cannibalise on their own dead.

挪威的老鼠喜欢吃肉类的残渣。热带地区的屋顶鼠喜欢吃水果。在城市地区，它们会吃宠物食品，扫荡垃圾桶，有的老鼠甚至会吃死去的同伴。

认知源于头脑对数据、信息、知识、理解和智慧的处理加工。信息是相关数据的组合。有用的信息可以提供知识。智慧是知识和深刻见解的升华。

Perceptions are processed in the mind as data, information, knowledge, understanding and wisdom. Information is data with relations. Information that is useful offers knowledge. Out of knowledge and insight comes wisdom.

数据和事实之间是有差异的。数据是静态信息。数据经筛选和过滤转换成加工后的信息，被称为事实。虽然数据是可以解释的，但事实就是事实，不能被解释。

There is a difference between data and fact. Data is a static piece of information. Sifted and filtered data turns into processed information that is referred to as fact. While data is open to interpretation, fact is fact and is not open to interpretation.

Think About It

想一想

Do you like having rats and cockroaches living around the house?

你喜欢住在周围有老鼠和蟑螂的房子里吗?

When Mum or Dad comes home with lots of food, are you happy or sad?

当爸爸妈妈带着很多食物回家时，你是快乐还是悲伤?

How humane is society when we have so much food waste that rats and cockroaches thrive?

人们浪费了那么多的食物，却让老鼠和蟑螂茁壮成长，这样的社会有多人道呢?

Why do you think the human brain is shrinking?

为什么你认为人类的大脑正在萎缩?

Do It Yourself!
自己动手!

Have a good look around your house, and in the area where you live. Are there any cockroaches? Have you ever seen any rats here? Talk to your parents and find out if, over the years, more or less of these little scavengers have been living off waste that people leave lying around. Ask people if they like rats, and also if they have ever killed cockroaches. Establish their methods of extermination and see if people are using toxic substances for this. Engage in a dialogue with friends and neighbours that have pets, and ask them what would happen to those pets when they eat a poisoned rat. Based on this, form you own opinion on safe ways of controlling rodents and insects in your homes and community. Now share it with others.

在你的房子周围和你居住的地方好好看看。有蟑螂吗？你在这儿见过老鼠吗？跟你的父母谈谈，这些年来，这些小食腐动物们是否或多或少都是靠人们的浪费而生存着。问问人们是否喜欢老鼠，还有他们是否曾经杀过蟑螂。验证他们所采取的灭绝蟑螂的方法，看看人们是否使用有毒物质。与有宠物的朋友和邻居聊聊，问问他们如果他们的宠物吃了中毒的老鼠会发生什么。在此基础上，形成你自己的意见并和别人分享：如何使用安全的方式控制你的家园和社区周围的啮齿动物和昆虫。

学科知识
Academic Knowledge

生物学	食物对大脑的影响；脑和脊髓是神经系统的中枢部分；大脑分为前脑、中脑和后脑。
化 学	神经化学催产素影响母性行为，谷氨酸或多巴胺是神经递质，血清素用来调节情绪和睡眠；肽和小蛋白支持神经元的生长和分化。
物 理	神经元是人脑基本的信号处理单位；神经网络。
工程学	神经想象技术。
经济学	心理学和神经科学发现决策受潜意识和情绪水平影响，会影响市场上许多商品的价格波动，由于价格的波动升值风险，经济理论认为价格在期望函数区间内具有不稳定性。
伦理学	人类具有分享的能力，并因分享感到快乐；作为一个物种的我们是多么富有同情心呢？
历 史	艾伦·劳埃德·霍奇金、安德鲁·赫胥黎和约翰·卡鲁·埃克尔斯因发现在神经细胞膜的外围和中心部位与神经兴奋和抑制有关的离子机理，获得1963年度诺贝尔生理学或医学奖。
地 理	卡拉哈里沙漠属非洲南部内陆干燥区，横跨纳米比亚、博茨瓦纳和南非。
数 学	用统计力学分析来自神经元的数据。
生活方式	过去一万年来，有限的营养导致了大脑的萎缩。
社会学	消除病虫害防治问题的根源有两种方式：规范和禁止，改变框架条件（清洁卫生）；海马体负责短时记忆的存储转换和定向等功能，在对短时记忆进行巩固进而转换成长期记忆中起着重要作用。
心理学	行为经济学家认为，投资者的消费心理对经济模式的发展有很大的影响；保护自我的乐观偏见导致高估的控制程度和成功概率。
系统论	决策是主观的，但许多决策的组合改变了市场，从而改变了看法，影响了现实；学习是通过改变突触的数量和强度以改变大脑生理的和社会化的过程。

情感智慧
Emotional Intelligence

蟑螂

蟑螂意识到他是在入侵老鼠的领地，所以他很有礼貌并要求获得许可。他感到惊讶的是，老鼠让他自己清理，因为老鼠认为蟑螂的角色是一个清洁工。蟑螂感到失望的是，当他试图做好事时，人们总想伤害他们。这让他讽刺并质疑人的友善，指出对人们要消灭蟑螂的不理解。他渴望了解人脑发生了什么以及人类的态度是如何变化的。他开始反思这种新的现实如何改变了人生的困境。这引发了一场积极的哲学讨论。

老鼠

老鼠表示很欢迎蟑螂，但他很快就说出了他欢迎来访者的条件。他对人类行为持批评态度，指出其多变性，比如人们试图消灭那些把他们留下的烂摊子清理干净的生物。他有一个积极的心态，特别是剖析了人们在过去的亿万年中的变化。老鼠分析了人类行为改变的原因，得出结论：当人失去积极情绪而充满消极意见时，大脑就会停止生长。他憧憬人类会变得更加人性化，期待所有生物都有一个更好的未来。

艺术
The Arts

表情符号给我们的大脑提供带有情感的符号印象，即使这个符号，如"：-）"只是代表了一张笑脸，我们也能够对这3个标志有一个积极的形象的解释。有些人可以看到月亮上的人、云上的鳄鱼，或山脊上的人的轮廓。所以你的任务是画一些抽象的东西。现在问一问你的朋友和家人从你的画中看到了什么。对他们可能看到的事物敞开大门，即使它不是你想要的那样。玩转颜色和对比度，把你的图像颠倒过来看，使用不同的光线照明，现在看看人们是如何反应的，这很有趣。

思维拓展
Systems: Making the Connections

随着时间的推移，人类的大脑发生了重大变化。首先是数百万年来，大脑的体积越来越大，使得人类能够结合情感和生存的需要，应付日益增长的感知和分析需求。科学家认为现在大脑的扩张是饮食变化的结果，也是生活方式的一个根本转变。然而在过去的一万年间，大脑已经开始萎缩，部分人认为这是一个暂时的现象。实际上我们已经从根本上改变了我们的饮食，从一个非常丰富和多样化的饮食结构，变成了以谷物、牛奶和肉类为主。由于新鲜时令水果、浆果、蔬菜，尤其是海鲜的新陈代谢，许多以前我们吸收的微量营养素消失了。社会随之进化，从前三到四代人居住在一个公共空间，现在越来越多的人选择独居的生活方式。虽然有许多需要考虑的假设，事实仍然是：我们目前的生活方式缺少了我们曾经拥有的警觉性和挑战。在某些方面，我们似乎生活得很富有——至少被日益增长的、威胁着我们的健康和生存的老鼠和蟑螂所赏识。我们中有多少人意识到，这些生物的扩散是人类不注意卫生以及浪费的生活方式造成的？我们需要转变我们的生活方式。随着越来越多的负面消息出现，数据表明我们当下的生活方式有很大问题，个人和社区正变得不堪重负，情况不容乐观。我们要改变自己的行为，清理自己的环境，而不是使用毒药，这不仅会杀灭啮齿动物和昆虫，也影响了我们的宠物，甚至让我们的孩子处于危险之中。所以问题是：我们为什么不准备改变？我们所在的地球的一切都在改变，如果我们不改变自己的思维和生活方式，世界就可能会改变我们。

动手能力
Capacity to Implement

身处社会中，人类怎样才能更人性化？我们想要快乐，也希望周围的人能过上幸福舒适的生活。那我们如何做到这一点？也许首先要了解人们想要什么。一旦我们想通了这一点，看看我们自己可以做些什么。去敲你邻居的门，开始交谈。这很可能会让邻居感到吃惊，但不要害羞或害怕，你的目标很清楚，是要给你的邻居（他或她）从来没有预料到的惊喜和幸福感。

故事灵感来自
This Fable Is Inspired by

莱斯·伯格
Lasse Berg

莱斯·伯格是一名作家和记者，1943出生于瑞典松兹瓦尔。他学过经济学和哲学，最初在亚洲当记者。20世纪80年代初，他移居非洲，居住在肯尼亚和埃塞俄比亚。他的研究涉及了与人类起源有关的一系列广泛课题。他详细分析了人类如何随着时代的变迁变得更加人性化的进程。

他的书《卡拉哈里沙漠的黎明：人类如何成为人类》为人们提供了社会如何从早期发展到现在的新见解。他的工作获得了许多奖项，他在质疑目前的智慧、推动科学走向新的水平、了解人类的起源、抓住我们的社会的潜力使之更人性化等方面的能力受到赞赏。

图书在版编目（CIP）数据

冈特生态童书.第四辑:修订版:全36册:汉英对照 /
(比)冈特·鲍利著;(哥伦)凯瑟琳娜·巴赫绘;
何家振等译.—上海:上海远东出版社,2023
书名原文:Gunter's Fables
ISBN 978-7-5476-1931-5

Ⅰ.①冈… Ⅱ.①冈…②凯…③何… Ⅲ.①生态环
境-环境保护-儿童读物—汉、英 Ⅳ.①X171.1-49

中国国家版本馆CIP数据核字(2023)第120983号
著作权合同登记号图字09-2023-0612号

策　　划　张　蓉
责任编辑　曹　茜
封面设计　魏　来　李　廉

冈特生态童书

人类能更人性化吗？

[比]冈特·鲍利　著
[哥伦]凯瑟琳娜·巴赫　绘

高　青　译

记得要和身边的小朋友分享环保知识哦！
八喜冰淇淋祝你成为环保小使者！